YOUR KNOWLEDGE HAS VALUE

Bradley Tice

Compressed Data for the Movie Industry

GRIN Verlag

Bibliografische Information der Deutschen Nationalbibliothek:

Die Deutsche Bibliothek verzeichnet diese Publikation in der Deutschen National-
bibliografie; detaillierte bibliografische Daten sind im Internet über http://dnb.d-
nb.de/ abrufbar.

Imprint:

Copyright © 2014 GRIN Verlag GmbH
Druck und Bindung: Books on Demand GmbH, Norderstedt Germany
ISBN: 978-3-656-59145-0

This book at GRIN:

http://www.grin.com/en/e-book/268095/compressed-data-for-the-movie-industry

GRIN - Your knowledge has value

Der GRIN Verlag publiziert seit 1998 wissenschaftliche Arbeiten von Studenten, Hochschullehrern und anderen Akademikern als eBook und gedrucktes Buch. Die Verlagswebsite www.grin.com ist die ideale Plattform zur Veröffentlichung von Hausarbeiten, Abschlussarbeiten, wissenschaftlichen Aufsätzen, Dissertationen und Fachbüchern.

Visit us on the internet:

http://www.grin.com/

http://www.facebook.com/grincom

http://www.twitter.com/grin_com

Compressed Data for the Movie Industry

Dr. Bradley S. Tice

Advanced Human Design

ABSTRACT

The paper will present a compression algorithm that will allow for both random and non-random sequential binary strings of data to be compressed for storage and transmission of media information. The compression system has direct applications to the storage and transmission of digital media such as movies, television, audio signals and other visual and auditory signals needed for engineering practicalities in such industries.

Introduction

The algorithmic compression program addressed in this paper was discovered by the author in 1998 as the only known algorithm that could 'compress' a random binary sequential string [1]. The key features of this algorithmic program are that they can be used on both random and non-random sequential strings of binary, and larger radix base numbers, and that both universal, whole, and specific, partial, aspects of compression of a linear sequential string can be utilized [2].

Foundations

The foundational aspects of this algorithm are that it treats all aspects of a linear sequential string as a concatenation of symbols that are based on the relevancy of the type of symbol, the placement of that symbol, upon a linear sequential string, and the adjoining symbols that my come before and after that specific symbol [3]. In other words, the environment that each and every specific symbol is upon a linear sequential string. In literature, the notion of a non-random sequential binary string is as follows:

Non-random binary sequential string: 1010101010

A random sequential binary string is as follows:

Random binary sequential string: 11101000011001111100

Notice that the non-random sequential binary string has a pattern of 'regularity' to it that has a 1 in the initial position followed by 0's and 1's for a collective of alternating five 1's and five 0's for a total of 10 characters and is able to be compressed and de-compressed to its original form [4].

The random sequential binary string has a 'less' harmonious, or balanced, pattern of symbols and is considered by traditional literature on the subject as being unable to compress [5].

Foundations – Examples

The following example, Example A, will be used to show the compression of a non-random binary sequential string:

Example A

Non-random sequential binary string: [10101010101010101010] of a total length of 20 symbols.

A practical notation for compression of the 20 symbols is to have the initials of the two different symbols; [1] and [0], and a symbol for a multiple of those two symbols [x] and for the number of times both symbols; [1] and [0], are accounted for in their respective place in the collective whole of the string, ten times, ten [1's] and ten [0's]. This results in the following formula: 10x10.

This compression can be notated as the following: [10] or a two, 2, character length.

Note that the figures [1] and [0] are not numbers, quantities, but rather qualities of divergence, separate values based on symbolic rather than numerical value. In other words, a [1] is the 'symbol' one rather than the number 1 and the figure [0] is the symbol zero rather than the numerical value of 0.

The random binary sequential string is compressed in Example B;

Example B

Random binary sequential string: [11111000101100011000] of a total length of 20 symbols.

Example B can be grouped into subdivisions of the whole by taking each common cluster of common, or like-natured, symbols into sub-groups of the complete string.

Sub-groups of a random binary sequential string as found in Example B:

[11111]+[000]+[1]+[0]+[11]+[000]+[11]+[000]

The resulting eight sub-groups provide for specific compression of each sub-group as follows:

[1]x5+[0]x3+[1]+[0]+[1]x2+[0]x3+[1]x2+[0]x3

This compression can be notated as the following: [10101010] or an eight, 8, character length.

Applications to Visual and Auditory Signals

The author originally designed the compression algorithm for telecommunications and computing applications [6]. The same qualities that give maximal compression to data signals also has direct applications to visual and auditory signals used in the various media arts industry that transmit and store digital data. In other words, compressed data for the movie industry.

Existing compression coding systems, such as Huffman coding, are not optimal when symbol-by-symbol coding is not used for lossless data compression [7]. While Huffman and arithmetic coding compression is superior to universal coding compression, universal code compression is simple and quick compared to Huffman coding [8].

In some respects all these older compression systems have a common extraneous feature of a 'prefix' code the is part of the core data and 'added on' like an arcane 'punch card' computer program. While necessary for the operation of these compression code systems the development of a 'prefix' free operating system has been developed by the author in the form of a simple computer system and my compression algorithm program [9].

The computer system designed uses a simple computer system by Emil Post in his 1936 paper and is integrated into my algorithm compression program that then functions as a left to right reading input of both random and non-random sequential binary strings that can be transmitted or stored and has the added feature of not using a prefix coding system [10].

The technical developments presented show a greater compression factor over existing practices and the development of a novel 'hybrid' computer system using a 1930's computer design with a late 20th century compression algorithm have shown vital technical developments that can be applied in visual and audio data signal engineering.

Conclusions

The algorithmic compression program has over a decade long theory and application experience in the engineering aspects to signal data transmission and storage with direct applications to telecommunications and computing. The following features have been developed in this paper for use in visual and auditory signal engineering systems.

The first feature is compression of random sequences. This allows for maximal compression of a sequential string.

The second feature is both a universal and specific quality that gives greater utilization of compression techniques to a sequential string.

The third feature is a proven system of algorithmic compression operation on radix base number systems greater than the traditional binary system.

The fourth feature is application to a computer system that has this compression algorithm as a more precise and functional system and integrated into a reliable computer design by one of the founders of computing Emil Post.

The fifth feature is that both random and non-random sequential strings compress and de-compress to their original states with the use of this algorithmic compression system.

Summary

The paper has presented research and development for the practical implementation of maximal compression of data for transmission and storage of media information via visual and audio data signals. While I have focused on the 'movie' industry, the growing distribution and storage needs of all visual and auditory arts media is considered for application of this compression system.

This umbrella coverage of data and information types is because most industries use both electronic communications and computers/computing in the development, transport, storage and implementation of visual and auditory information and the algorithmic program discussed in this paper has a wide and utilitarian application to many aspects to the engineering portion of these processes.

References

[1]. Tice, B.S. (2012) <u>A Level of Martin-Lof Randomness.</u>

New Hampshire: Science Publishers/CRC Press.

[2]. Tice (2012), abide, p. 34-35.

[3]. Tice (2012), abide.

[4]. Tice (2012), abide.

[5]. Tice (2012), abide,

[6]. Tice, B.S. (2009) <u>Aspects of Kolmogorov Complexity: The Physics of Information.</u>

Denmark: River Publishers, page Appendix section.

[7]. Wikipedia (2013a) "Huffman coding". Wikipedia Encyclopedia, November 27, 2013, pp. 1-10.

Website: http://en.wikipedia.org/wiki/Huffman_coding page 1.

[8]. Wikipedia (2013b) "Universal code (Data compression)". Wikipedia Encyclopedia, November 27,

2013, pp. 1-3. Website: http://en.wikipedia.org/wiki/Universal_code_(data_compression) page 2.

[9]. Tice (2012b) "A Universal Archetype Computer System". Munich, Germany: GRIN Verlag.

[10]. Tice, (2012b), abide.